Firefighter Fatalities in the United States in 1997

Prepared for

United States Fire Administration
Federal Emergency Management Agency
Contract No. EME-98-SA-0083

Prepared by

IOCAD Emergency Services Group
TriData Corporation

August 1998

Table of Contents

ACKNOWLEDGMENTS

This study of firefighter fatalities would not have been possible without the cooperation and assistance of many members of the fire service across the United States. Members of individual fire departments, chief fire officers, the National Interagency Fire Center, US Forest Service personnel, the US military, the Department of Justice, and many others contributed important information for this report.

IOCAD Emergency Services Group of Emmitsburg, Maryland (a division of IOCAD Engineering Services, Inc.), in conjunction with TriData Corporation of Arlington, Virginia conducted this analysis for the United States Fire Administration under contract EME-98-SA-0083.

The ultimate objective of this effort is to reduce the number of firefighter deaths through an increased awareness and understanding of their causes and how they can be prevented. Fire fighting, rescue, and other types of emergency operations are essential activities in an inherently dangerous profession, and unfortunate tragedies occur. This is the risk all firefighters accept every time they respond to an emergency incident. However, the risk can be greatly reduced through efforts to increase firefighter health and safety.

The United States Fire Administration would like to extend its thanks to Jack Jordan of the Phoenix, Arizona, Fire Department for providing the photograph for the cover.

This report is dedicated to the families of those firefighters who made the ultimate sacrifice in 1997. May the lessons learned from their passing not go unheeded.

BACKGROUND

For over 20 years, the United States Fire Administration (USFA) has tracked the number of firefighter fatalities and conducted an annual analysis. Through the collection of information on the causes of firefighter deaths, the USFA is able to focus on specific problems and direct efforts towards finding solutions to reduce the number of firefighter fatalities in the future. This information is also used to measure the effectiveness of current programs directed toward firefighter health and safety.

In addition to the analysis, the USFA maintains a list of firefighter fatalities for the Fallen Firefighters Memorial Service. The fallen firefighter's next of kin, as well as members of the individual fire department, are invited to the annual Fallen Firefighters Memorial Service, which is held at the National Fire Academy in Emmitsburg, Maryland every fall. Additional information regarding the memorial service can be found on the internet at **http://www.usfa.fema.gov/ffmem/service.htm** or by calling the National Fallen Firefighters Foundation at (301)-447-1365. An updated list of firefighter fatalities from 1981 through the present can be found at **http://www.usfa.fema.gov/ffmem/roll.htm**

INTRODUCTION

This report continues a series of annual studies by the United States Fire Administration of on-duty firefighter fatalities in the United States.

The specific objective of this study was to identify all of the on-duty firefighter fatalities that occurred in the United States in 1997, and to analyze the circumstances surrounding each occurrence. The study is intended to help identify approaches that could reduce the number of firefighter deaths in future years.

In addition to the 1997 overall findings, this study includes special analyses on collisions involving vehicles operated by firefighters and fatalities that occurred when firefighters working at the scene of an emergency were struck and killed by civilian vehicles.

Who Is a Firefighter?

For the purpose of this study, the term firefighter covers all members of organized fire departments, including career and volunteer firefighters; full-time public safety officers acting as firefighters; state and federal government fire service personnel; including wildland firefighters; and privately employed firefighters, including employees of contract fire departments and trained members of industrial fire brigades, whether full or part-time. It also includes contract personnel working as firefighters or assigned to work in direct support of fire service organizations.

Under this definition, the study includes not only local and municipal firefighters, but also seasonal and full-time employees of the United States Forest Service, the Bureau of Land Management, the Bureau of Indian Affairs, the Bureau of Fish and Wildlife, the National Park Service, and state wildland agencies. The definition also includes prison inmates serving on fire fighting crews; firefighters employed by other governmental agencies such as the United States Department of Energy; military personnel performing assigned fire suppression activities; and civilian firefighters working at military installations.

What Constitutes an On-Duty Fatality?

On-duty fatalities include any injury or illness sustained while on-duty that proves fatal. The term on-duty refers to being involved in operations at the scene of an emergency, whether it is a fire or non-fire incident; being en route to or returning from an incident; performing other officially assigned duties such as training, maintenance, public education, inspection, investigations, court testimony and fund-raising; and being on-call, under orders, or on stand-by duty, except at the individual's home or place of business.

These fatalities may occur on the fireground, in training, while responding to or returning from alarms, or while performing other duties that support fire service operations.

A fatality may be caused directly by accident or injury, or it may be attributed to an occupational-related fatal illness. A common example of a fatal illness incurred on duty is a heart attack. Fatalities attributed to occupational illnesses would also include a communicable disease contracted while on duty that proved fatal, where the disease could be attributed to a documented occupational exposure.

Accidents that claim the lives of on-duty firefighters are also included in the analysis, whether or not they are directly related to emergency incidents. For example, in 1997, this category includes a firefighter that was killed when an open propane tank exploded in the station while he was on duty.

Injuries and illnesses are included where death is considerably delayed after the original incident. When the incident and the death occur in different years, the analysis counts the fatality as having occurred in the year that the incident occurred. Two firefighters died in 1997 as the result of injuries and exposures they suffered in previous years. One firefighter died of medical complications that resulted from an ambulance collision that occurred in 1996 and another died in from complications of AIDS contracted through documented needle sticks while on the job as a paramedic. Because these deaths were the result of incidents that occurred prior to 1997, these cases are counted as 1996 and 1980 fatalities, respectively, for statistical purposes. They are not included in the 94 fatalities for 1997 that were analyzed in this report. Since these deaths occurred in 1997, these two firefighters will be included in the 1997 annual Fallen Firefighters Memorial Service at the National Fire Academy, and their names will be included on the list of firefighters who died in 1997.

There is no established mechanism for identifying fatalities that result from illnesses that develop over long periods of time, such as cancer, which may be related to occupational exposure to hazardous materials or products of combustion. It has proven to be very difficult over several years to provide a full evaluation of an occupational illness as a causal factor in firefighter deaths, because of the limitations in the ability to track the exposure of firefighters to toxic hazards, the often delayed long-term effects of such exposures, and the exposures firefighters may receive while off-duty.

Sources of Initial Notification

As an integral part of its ongoing program to collect and analyze fire data, the United States Fire Administration solicits information on firefighter fatalities directly from the fire service and from a wide range of other sources. These sources include the Public Safety Officer's Benefit Program (PSOB) administered by the Department of Justice, the Occupational Safety and Health Administration (OSHA), the US military, the National Interagency Fire Center, and other federal agencies.

The USFA receives notification of some deaths directly from fire departments, as well as from fire service organizations such as the International Association of Fire Chiefs (IAFC), the International Association of Fire Fighters (IAFF), the National Fire Protection Association (NFPA), the National Volunteer Fire Council (NVFC), state fire marshals, state training organizations, other state and local organizations, and fire service publications. The USFA also keeps track of fatal fire incidents as part of its Major Fire Investigations Project and maintains an ongoing analysis of data from the National Fire Incident Reporting System (NFIRS) for the production of the report "Fire in the United States".

Procedure for Including a Fatality in the Study

In most cases, after notification of a fatal incident, initial telephone contact is made with local authorities by the USFA's contractor to verify the incident, its location and jurisdiction, and the fire department or agency involved. Further information about the deceased firefighter and the incident may be obtained from the Chief of the fire department or his or her designee over the phone or by other data collection forms.

Information that is routinely requested includes NFIRS-1 (incident) and NFIRS-3 (fire service casualty) reports, the fire department's own incident reports and internal investigation reports, copies of death certificates or autopsy results, special investigative reports such as those produced by the USFA or NFPA, police reports, photographs and diagrams, and newspaper or media accounts of the incident.

After obtaining this information, a determination is made as to whether the death qualifies as an on-duty firefighter fatality according to the previously described criteria. The same criteria were used for this study as in previous annual studies. Additional information may be requested, either by follow-up with the fire department directly, from state vital records offices, or other agencies. The determination as to whether a fatality qualifies as an on-duty death for inclusion in this statistical analysis is made by the USFA. The final determination as to whether a fatality qualifies as a line of duty death for inclusion in the National Fallen Firefighters Memorial Service is made by the National Fallen Firefighters Foundation.

1997 FINDINGS

Ninety-four (94) firefighters died while on duty in 1997.[1] This represents a drop of one death from 1996, and is slightly lower than the 96 firefighters who gave their lives in 1995. The total of 94 fatalities is the third lowest number recorded in the 20 years that this data has been collected, and is only the fifth time that the total has been less than 100 fatalities. The lowest years were 1992, with 75 fatalities and 1993, with 77 fatalities.

This year's total continues the long-term downward trend of reduced fatalities that began in 1979, after a peak of 171 in 1978. The overall trend in firefighter fatalities is down twenty percent over the last ten years. However, the rate of reduction in the last five years has slowed to four percent, partly due to the uncharacteristically low number of deaths that occurred in 1992 and 1993 (Figure 1).

Figure 1 - On-Duty Firefighter Fatalities (1977-1997)

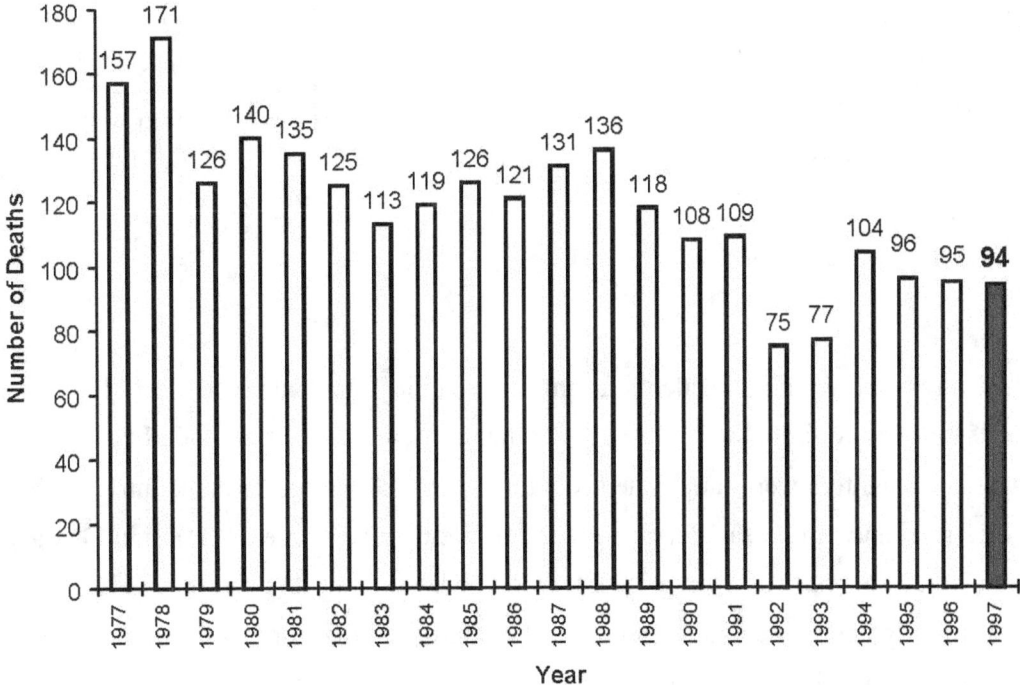

---

[1] As mentioned earlier, the 94 on-duty fatalities in 1997 do not include two firefighters who died in 1997 as the result of injuries or exposures received in prior years.

1997 firefighter fatalities included 57 volunteer firefighters and 37 career firefighters (up from 26 career in 1996) (Figure 2). Among the volunteer firefighter fatalities, 55 were from local or municipal volunteer fire departments and two were seasonal or contract members of wildland fire agencies. Of the career firefighters who died, 32 were members of local or municipal fire departments and five were wildland firefighters. Ninety-three of the fatalities were men and one was a woman.

Figure 2 - Career, Volunteer, and Wildland Fatalities (1997)

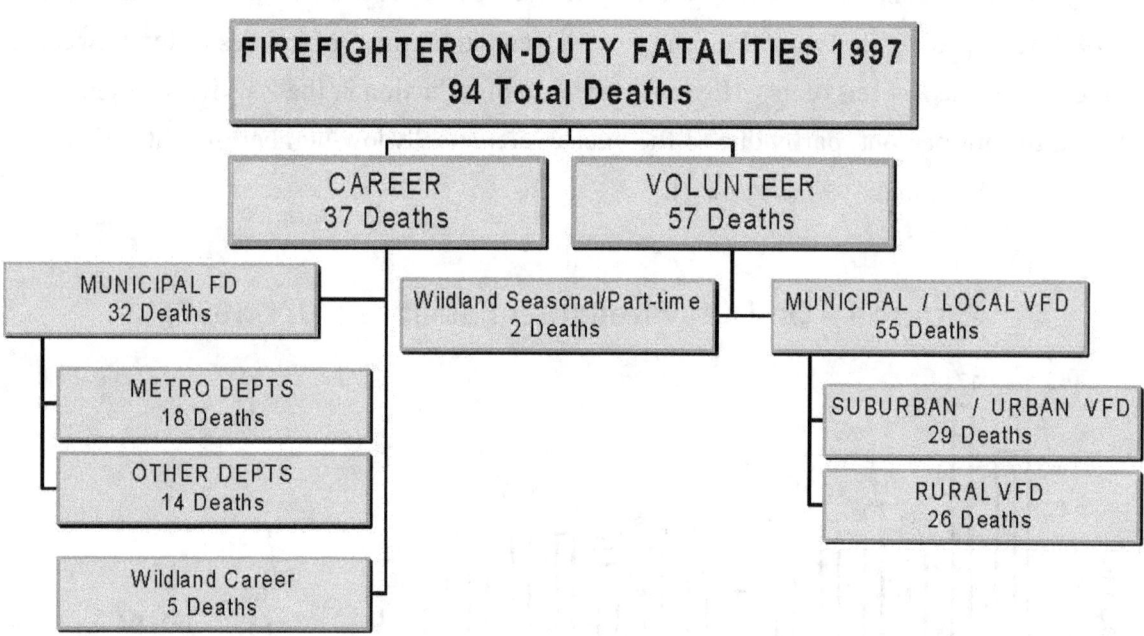

The 94 deaths resulted from 85 incidents. Eight multi-fatality incidents resulted in 17 firefighter deaths. Three firefighters were killed when a chemical plant exploded in Arkansas; two firefighters were fatally electrocuted when one firefighter came into contact with live electrical lines and created a chain reaction during a vehicle extrication in Ohio; two were killed as the result of a large LPG tank explosion at a grain dryer fire in Illinois; two Missouri firefighters were killed responding to a motor vehicle collision when a dump truck crossed the center line and struck their fire apparatus; two flight crews of two members each were killed in wildland aircraft crashes in Arizona and Pennsylvania; two firefighters died in a collapse during a structure fire in California, and two Pennsylvania firefighters died in a residential structure fire.

The number of deaths associated with brush, grass or wildland fire fighting rose to nine from the five deaths experienced in 1996. These two years reflect a significant drop from the 18 firefighters that died in wildland activities in 1995. Five firefighters were killed in three wildland fire fighting aircraft crashes in Arizona, California, and Pennsylvania; one died from a severe asthma attack at a prescribed burn in Nevada; one firefighter died of Strep A infection contracted while fighting a wildland fire in California; a firefighter in Florida died of exposure; and a firefighter in California died of heat stroke.

Type of Duty

In 1997, 76 firefighter on-duty deaths were associated with emergency incidents, accounting for 81 percent of the 94 fatalities (Figure 3). This includes all firefighters who died while responding to an emergency, while at the emergency scene, or after the emergency incident. Non-emergency activities accounted for 18 fatalities (19 percent). Non-emergency duties include training, administrative activities, or performing other functions that are not related to an emergency incident. Five firefighters died while participating in training exercises, including one who was killed when an aerial device was retracted during training and one that was killed during a dive rescue training session. One firefighter was killed when an open propane cylinder was rolled into a rescue station as a practical joke and exploded; two firefighters died at official meetings and one at a memorial service; one died during a prescribed burn; and six died of heart attacks or strokes while on-duty.

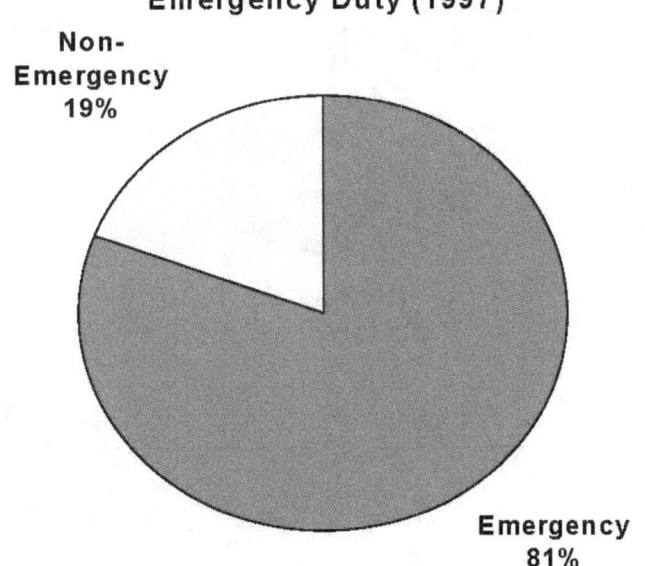

Figure 3 - Firefighter Deaths While Performing Emergency Duty (1997)

Non-Emergency 19%

Emergency 81%

The number of deaths by type of duty being performed is shown in Table 1 and presented graphically in Figure 4. As in previous years, the largest number of deaths occurred during fireground operations. There were 41 fireground deaths, which accounted for 43.6 percent of the fatalities, up over three percent from 1996. Of the 41 fireground deaths, over one-third (14) resulted from heart attacks that occurred on the fire scene. Other fireground deaths included 11 from asphyxiation, 10 from internal trauma, four from burns, one from heat stroke, and one from Strep A.

Table 1. Type of Duty – 1997	Number	Percent
Fireground Operations	41	43.6%
Responding / Returning from Alarm	20	21.3%
Other On-Duty	15	16.0%
Non-Fire Emergencies	13	13.8%
Training	5	5.3%
TOTAL	94	100%

Figure 4 - Fatalities by Type of Duty (1997)

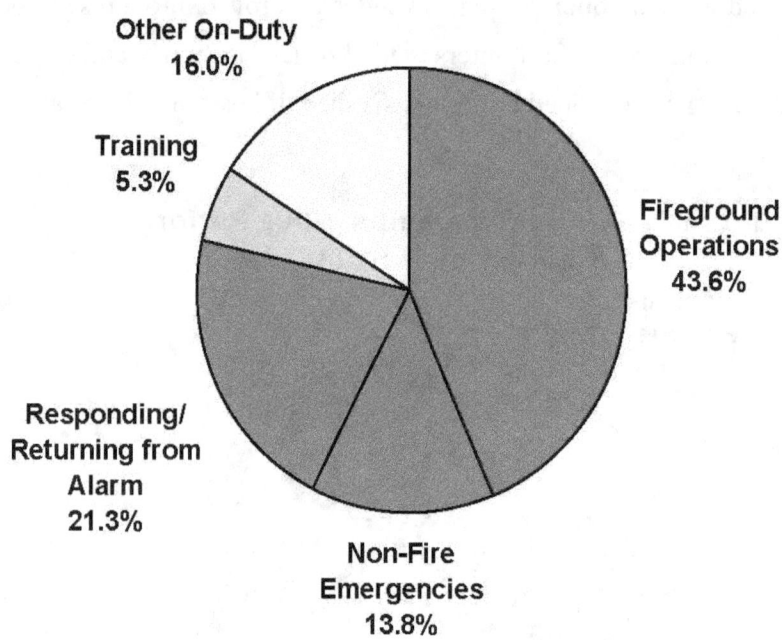

The second largest category of deaths by duty type was responding to or returning from emergency incidents, which accounted for 20 deaths in 1997 (down two deaths from 1996). This has been the second leading type of duty in which firefighter deaths have occurred each year since 1993. Seventeen of the deaths involved volunteer firefighters. Five firefighters suffered fatal heart attacks while responding to or returning from emergency incidents. Fourteen firefighters were killed in fire apparatus accidents while enroute to emergency incidents, six of these deaths involved rollovers. Five firefighters were killed in accidents involving their personal vehicles while enroute to emergency calls. One of these involved a firefighter who was killed when a tree blown over by a storm crushed his car. His son, who was also a firefighter, was injured. Collisions involving vehicles operated by firefighters will be addressed in more detail later in this report.

Thirteen deaths were related to activities at the scene of non-fire emergency incidents. Four firefighters were killed in separate incidents when they were struck by passing vehicles while on the scene of non-fire emergencies; three drowned in separate rescue incidents; three died from heart attacks; two were electrocuted while engaged in a vehicle extrication, and one firefighter was killed when the ambulance in which he was riding was involved in a collision.

There were 15 deaths that occurred during non-emergency duty activities. These deaths include thirteen firefighters who died from heart attacks while on duty – six upon returning from an incident, two at fire department meetings, one at a fire department memorial service, and one during mandatory physical training (exercise).

Five deaths were attributed to training activities, including deaths during dive rescue training, aerial device operation training, and rookie school. Two training deaths were from heart attacks, one involving a student and one involving an instructor.

Career, Volunteer, and Wildland Deaths by Type of Duty

Figure 4a depicts career, volunteer, and wildland firefighter deaths by type of duty. Wildland career, wildland seasonal, and wildland contractor deaths were grouped together. This chart demonstrates the disproportionate number of fatalities experienced by volunteer firefighters responding to and returning from alarms when compared with career and wildland firefighters. This issue will be addressed in more detail in the special topics area of this report.

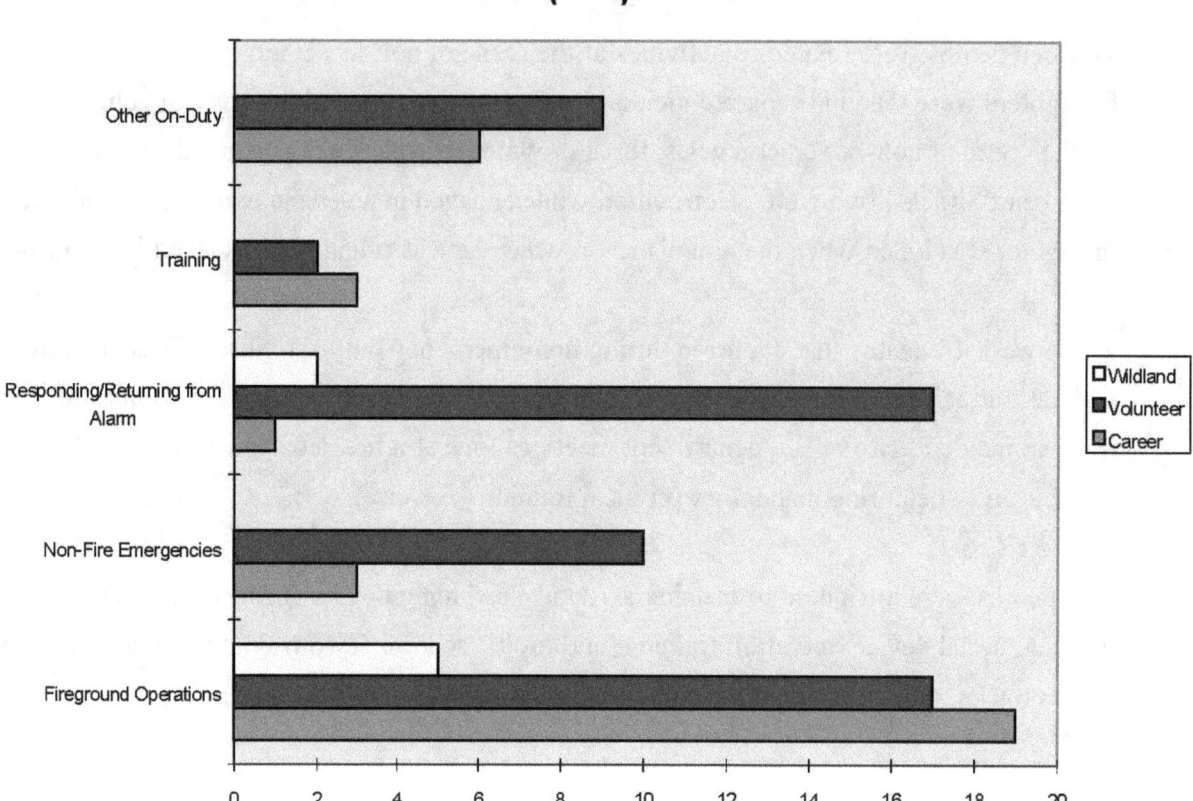

Figure 4a - Career, Volunteer, and Wildland Deaths by Type of Duty (1997)

Type of Emergency Duty

76 firefighters died while directly engaged in emergency service delivery, including deaths that were the result of injuries sustained on the incident scene or enroute to the incident scene. Figure 4b shows the percentage of firefighters killed in fire fighting, emergency medical services, hazardous materials, and technical rescue related incidents. 47 firefighters were killed in relation to fires, 20 were killed in relation to EMS calls, five were killed in association with hazardous materials emergencies, three were killed while engaged in technical rescues and the nature of the emergency for one firefighter was unknown.

Figure 4b - Type of Emergency Duty (1997)

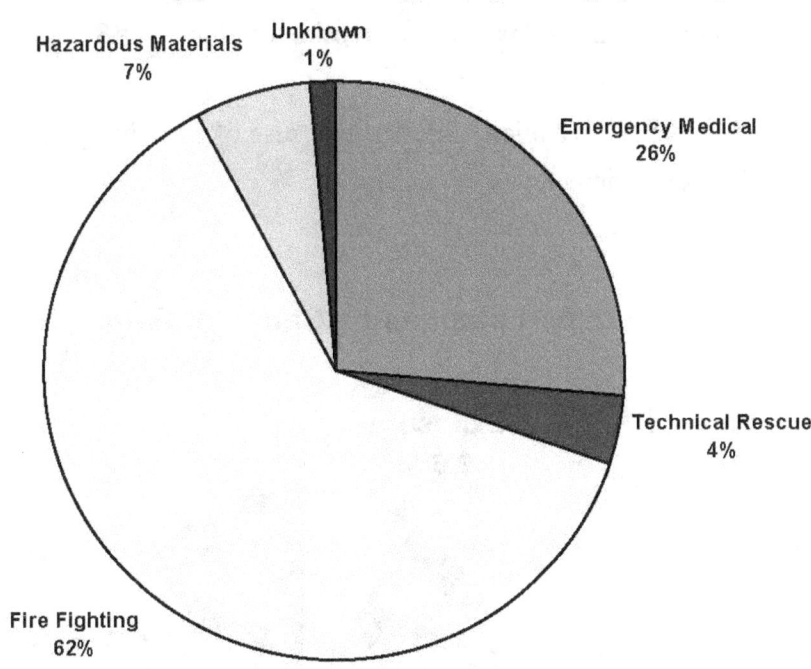

On-Scene or Enroute - 76 of 94

Cause of Fatal Injury

As used in this study, the term 'cause of injury' refers to the action, lack of action, or circumstances that directly resulted in the fatal injury, while the term nature of injury refers to the medical cause of the fatal injury or illness, often referred to as the physiological cause of death. A fatal injury usually is the result of a chain of events, the first of which is recorded as the cause. For example, if a firefighter is struck by a collapsing wall, becomes trapped in the debris, runs out of air before being rescued, and dies of asphyxiation, the cause of the fatal injury is recorded as "struck by collapsing wall" and the nature of the fatal injury is "asphyxiation". Similarly, if a wildland firefighter is overrun by a fire and dies of burns, the cause of the death would be listed as "caught/trapped," and the nature of death would be "burns". This follows the convention used in NFIRS casualty reports.

Figure 5 shows the distribution of deaths by cause of fatal injury or illness and Table 2 presents the exact number.

Figure 5 - Fatalities by Cause of Fatal Injury (1997)

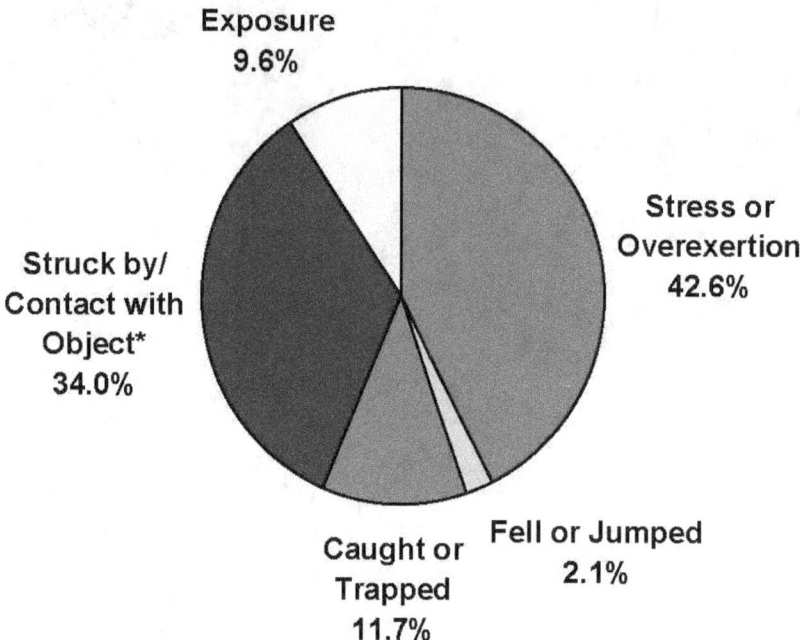

Table 2. Cause of Fatal Injury	Number	Percent
Stress or Overexertion	40	42.6%
Struck By/Contact with an Object	32	34.0%
Caught or Trapped	11	11.7%
Exposure	9	9.6%
Fell or Jumped	2	2.1%
TOTAL	94	100%

As in most previous years, the largest cause category is stress or overexertion, which was listed as the primary factor in 42.6 percent of the deaths, over six percent less than last year. Fire fighting is extremely strenuous physical work and is likely one of the most physically demanding activities that the human body performs. Most firefighter deaths attributed to stress result from heart attacks. Of the 40 stress-related fatalities in 1997, 36 firefighters died of heart attacks, two died of strokes (CVA's), and two died of heat stroke/exhaustion. Fifteen of the 40 deaths whose cause is listed as stress/exertion occurred during non-emergency activities.

The second leading cause of firefighter fatalities was being struck by or coming in contact with an object. Of the 32 firefighters who died in these incidents, 17 were involved in motor vehicle accidents, five were killed in aircraft crashes, three were killed in an explosion, two were crushed by trees, one was killed in aerial device training, and one was killed when struck by an overheated bullet associated with a greenhouse fire. Three others died in miscellaneous incidents.

The third leading cause of firefighter fatalities was being caught or trapped, which accounted for 11 deaths (11.7 percent), up more than four percent from 1996. Five firefighters died after becoming trapped by collapses, three were killed in basement fires, and three firefighters drowned - two during rescue attempts and one during training.

Nine deaths were attributed to exposure[2]. Two firefighters died after removing their SCBA in a structural fire, two were killed in an LPG explosion, one was killed in an explosion when another firefighter rolled an open propane cylinder into a rescue station as a practical joke, one died from Strep A, one died while fighting a restaurant fire, one died from

---

[2] "Exposure/Contact with" follows NFIRS 4.0 definitions under "Cause of Fatal Injury".

respiratory and cardiac arrest at a ten acre brush fire, and one firefighter died in association with a chemical exposure at a structure fire.

Two firefighters died as a result of falls. One firefighter was killed in a roof collapse as he was on a roof checking the progress of a fire and the other firefighter was thrown from a fire truck as it rolled enroute to a fire call.

Nature of Fatal Injury

Table 3 and Figure 6 show the distribution of the 94 deaths by the medical nature of the fatal injury or illness. The leading nature of death in 1997 was heart attacks, which accounted for 36 firefighter fatalities (ten less than 1996). Two firefighters were felled by strokes (CVA's), one of which occurred at the scene of a motor vehicle collision.

Table 3. Nature of Fatal Injury	Number	Percent
Heart Attacks	36	38.3%
Stroke/Seizure	2	2.1%
Internal Trauma	32	34.0%
Asphyxiation (includes drowning)	15	16.0%
Burns	4	4.3%
Electrocution	2	2.1%
Heat Exhaustion/Stroke	2	2.1%
Other	1	1.1%
TOTAL	94	100%

Figure 6 - Fatalities by Nature of Fatal Injury (1997)

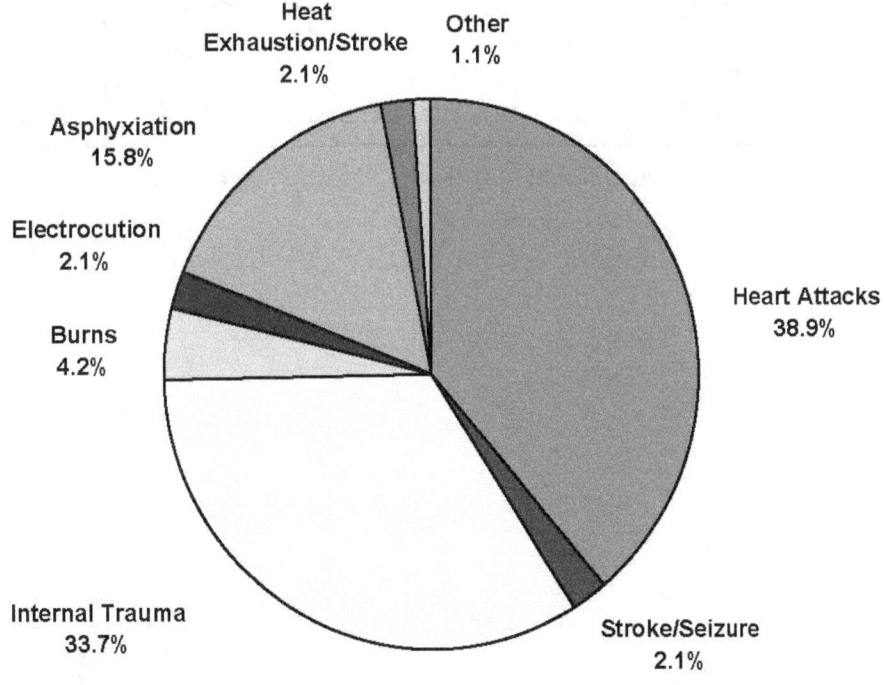

Figure 6a provides a detailed breakdown of heart attacks by type of duty. Fourteen of the heart attacks occurred at the fire scene and five occurred while enroute to or returning from an emergency incident, including one firefighter who collapsed while walking or running to the fire station for an emergency. Three occurred at training incidents and five occurred during EMS incidents. Ten heart attacks occurred during other on-duty situations A heart attack struck down one additional firefighter as he slept.

Figure 6a - Heart Attacks by Type of Duty

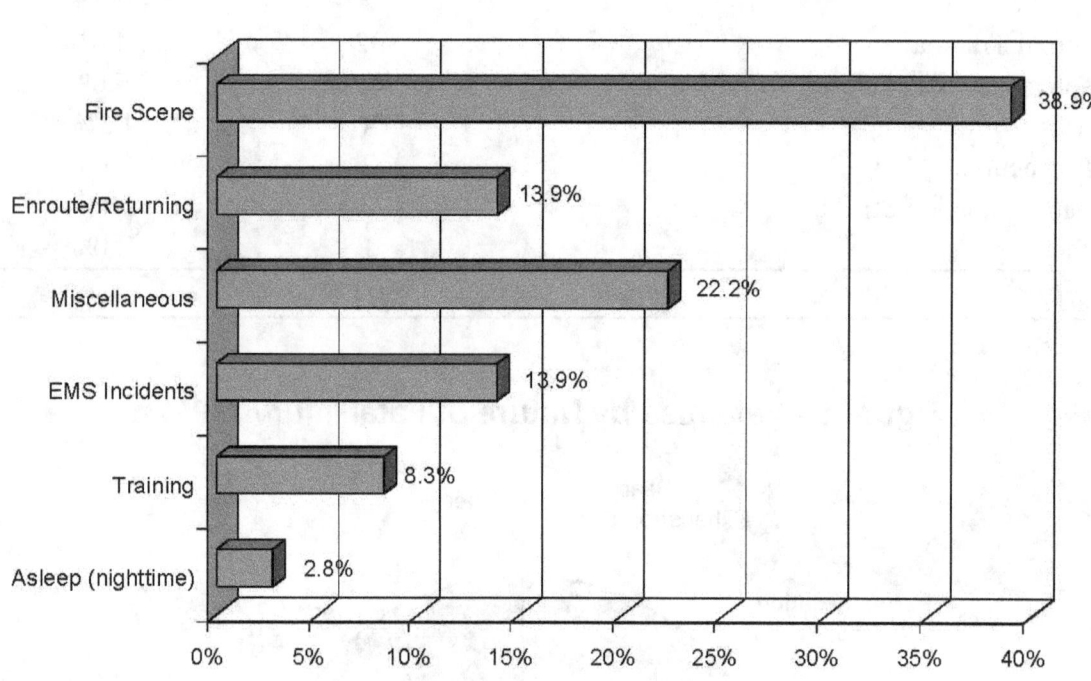

Internal trauma was the second leading nature of death, responsible for 32 deaths (the same number as the revised total for 1996). This total includes 14 firefighters who were involved in vehicle accidents, five who were hit by vehicles while on the emergency scene, five who died in wildland fire fighting aircraft crashes, three that were killed in an explosion, one who was killed by a falling tree, one who was killed as the result of a wall collapse, one who died during aerial device training, one who was killed by a bullet at a greenhouse fire, and one death occurred after fighting a tire fire.

Asphyxiation was the third leading medical reason for firefighter deaths, responsible for 15 deaths, up from five deaths in 1996. Ten of these deaths occurred during structural fire fighting, three were drownings during rescues or training, one occurred at a prescribed burn, and one was the result of a practical joke involving an explosion.

Four of the 94 firefighter fatalities that occurred in 1997 were attributed to burns. Two firefighters were killed in the explosion of a large LPG tank at a grain dryer fire and two firefighters were killed in separate structure fires.

Two firefighters died from electrocution. They were electrocuted at the scene of a motor vehicle collision when one rescuer came into contact with downed power lines, creating a chain reaction.

Two firefighters were felled by strokes (CVA's), one of which occurred at the scene of a motor vehicle collision..

The medical cause of death for one firefighter was Strep A. The firefighter was working a wildland fire when he became ill and was transported to a hospital where he was diagnosed with Strep A and died two days later. This disease has symptoms similar to the flu.

Firefighters Ages

Figure 7 shows the distribution of firefighter deaths by age and cause of death. Younger firefighters were more likely to have died as a result of traumatic injuries from an apparatus accident or after becoming caught or trapped during fire fighting operations. Stress was shown to play an increasing role in firefighter deaths as age increased.

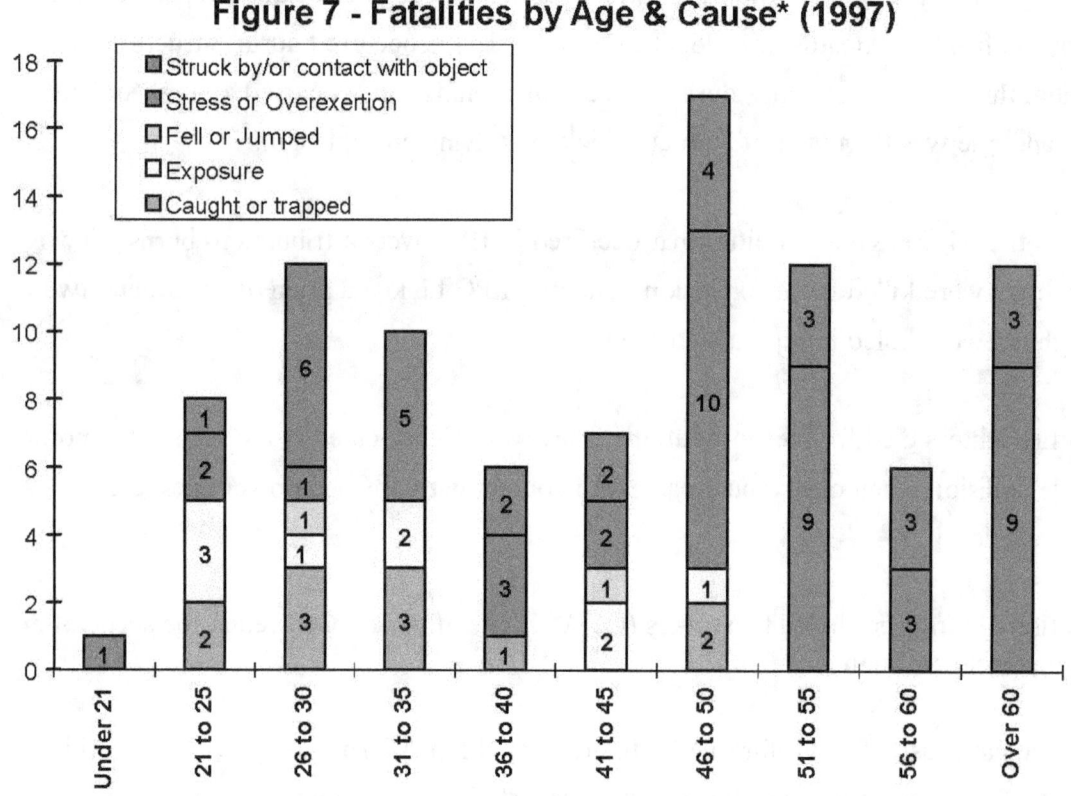

Figure 7 - Fatalities by Age & Cause* (1997)

* - 91 out of the 94 deaths reported age

This is also reflected in Figure 8, which shows the distribution of deaths by age and medical nature of injury. Trauma and asphyxiation were responsible for most of the deaths of younger firefighters, while heart attacks were much more prevalent among older firefighters. Heart attacks accounted for 66.6 percent of firefighter deaths where the firefighter was age 51 or higher.

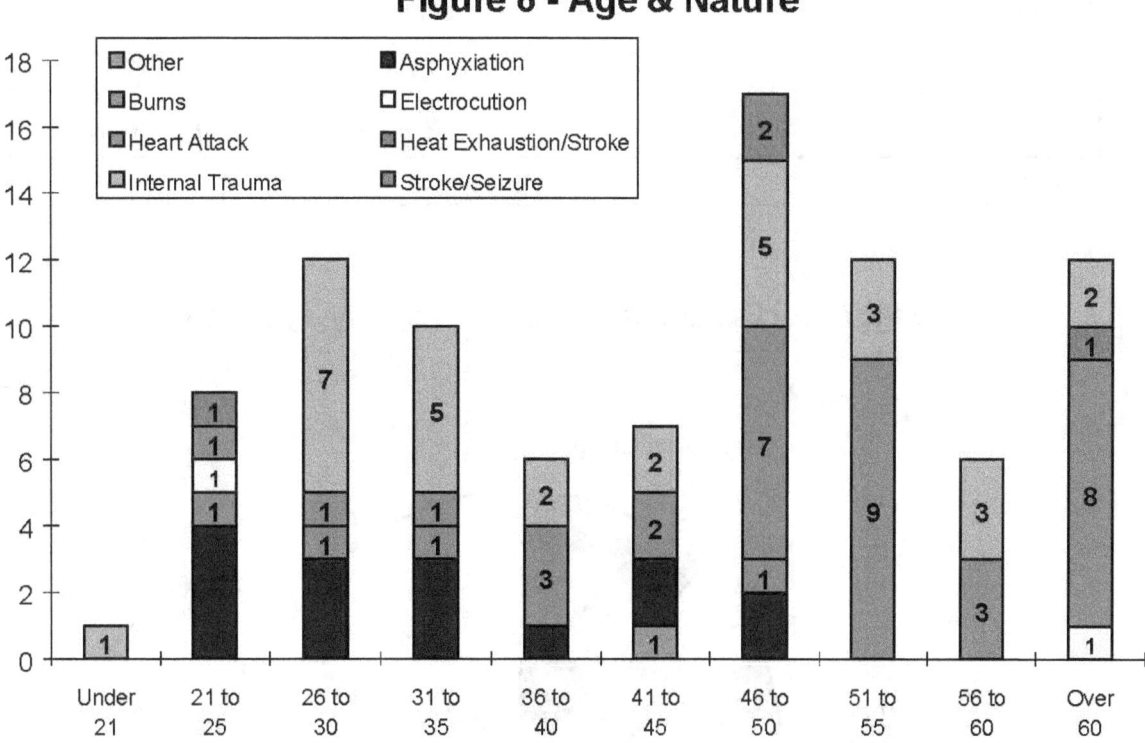

Figure 8 - Age & Nature*

* - 91 out of the 94 deaths reported age

Fixed Property Type

There were 41 fireground deaths in 1997, an increase of three from 1996. Figure 9 and Table 4 show the distribution by fixed property use.

Table 4. Property Use for Fireground Deaths	Number	Percent
Residential	16	39.0%
Commercial	11	26.8%
Outdoor Property	8	20.0%
Manufacturing	5	12.2%
Street / Road	1	2.4%
TOTAL	41	100%

Figure 9 - Fatalities by Fixed Property Type (1997)

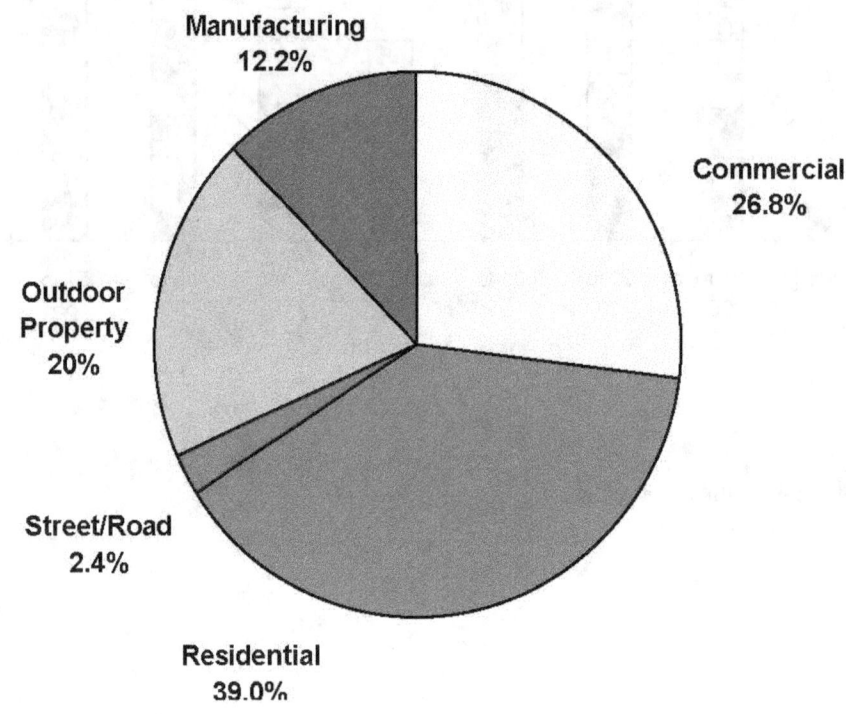

Fireground deaths only - 41 of 94

Fireground Deaths

Structural fires accounted for 32 fireground deaths. As in most years, residential occupancies accounted for the highest number of these fireground fatalities, with 16 deaths (half of all structural fire deaths). Residential occupancies usually account for 70-80 percent of all structure fires and a similar percentage of the civilian fire deaths each year[3]. The frequency of firefighter deaths in relation to the number of fires is much higher for non-residential structures.

Outdoor properties and "street/road" accounted for a total of nine deaths. Three deaths were the result of wildland fire fighting aircraft crashes, one from a fallen tree at a debris fire, one from heat stroke, one from Strep A, one from asthma, one from cardiac and respiratory arrest at a brush fire, and one from being struck by a motor vehicle at a car fire.

Type of Activity

Figure 10 and Table 5 show the type of fireground activity that the 41 firefighters were engaged in at the time they sustained their fatal injuries or illnesses.

Table 5. Type of Activity for Fireground Deaths	Number	Percent
Advancing Hose Lines / Fire Attack	21	51.2%
Search and Rescue	5	12.2%
Water Supply	4	9.8%
Support Duties	3	7.3%
Incident Command	3	7.3%
Cutting Fire Breaks (Wildland)	3	7.3%
Ventilation	2	4.9%
TOTAL	41	100%

---

[3] Complete NFIRS data for 1997 fire incidence was not available at the time of this report, but residential fires typically account for between 70 and 80 percent of all civilian fatalities each year.

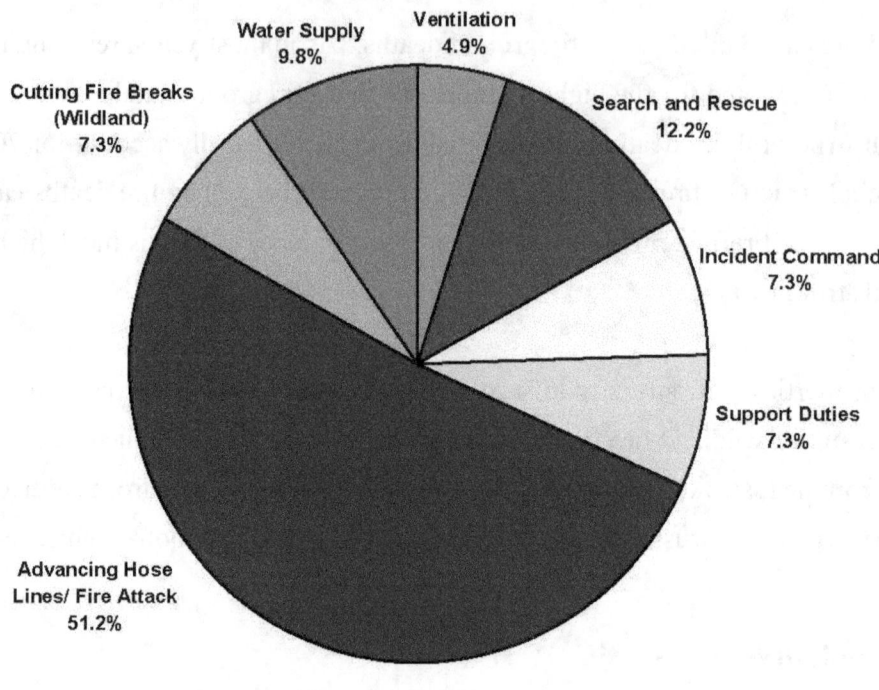

Figure 10 - Fatalities by Type of Activity (1997)

Fireground Operations Deaths Only - 41 of 94

Compared to 1996, there was a substantial increase this year in the number of firefighters who died while engaged in fire attack and advancing hose lines (increase of 12). The 1997 level is still below the 23 firefighters killed performing this activity in 1995. Twelve of these twenty-one firefighters were killed engaged in structural fire fighting, five were killed in explosions, three were killed in wildland fire fighting aircraft crashes, and one was killed by a fallen tree at an outdoor fire.

Five firefighters were killed during search and rescue operations (the same number were killed in 1996). Two were trapped in a collapse of a residential structure fire, two while fighting a residential fire, and one from exposure to smoke and carbon monoxide at a structural fire.

Four firefighters were killed while engaged in water supply operations (a drop of four from 1996). All were killed at structural fires.

Three firefighters became ill while cutting fire breaks, one from heat stroke, one from Strep A, and one from cardiac and respiratory arrest. Three firefighters died while performing support duties, all of heart attacks. Two firefighters died performing ventilation duties, one when a car struck him as he prepared to open a hood at a car fire and the other as the result of a roof collapse.

Three firefighters acting in incident command roles died in 1996, two of heart attacks and one of respiratory arrest leading to heart failure at a prescribed burn. One incident involved the death of a Chief officer during a mutual aid response to structure fire in a neighboring city.

Time of Alarm

The distribution of all 1997 firefighter deaths according to the time of day when the incidents were reported is shown in Figure 11 (46 times were not reported). For structural fire fighting deaths only, two thirds occur between 8:00 p.m. and 8:00 a.m.

Figure 11 - Fatalties by Time of Death* (1997)

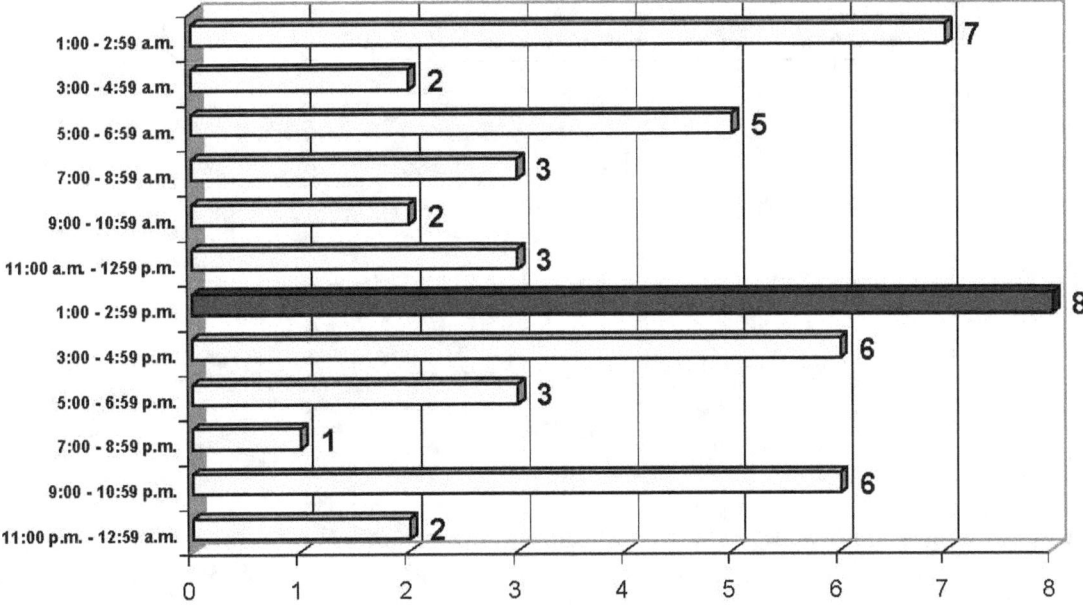

46 times were not reported

Month of the Year

Figure 12 illustrates firefighter fatalities by month of the year. Firefighter fatalities peaked in January and June. Other high months were recorded in May, October, and November. The spring months (March and April) were among the lowest months.

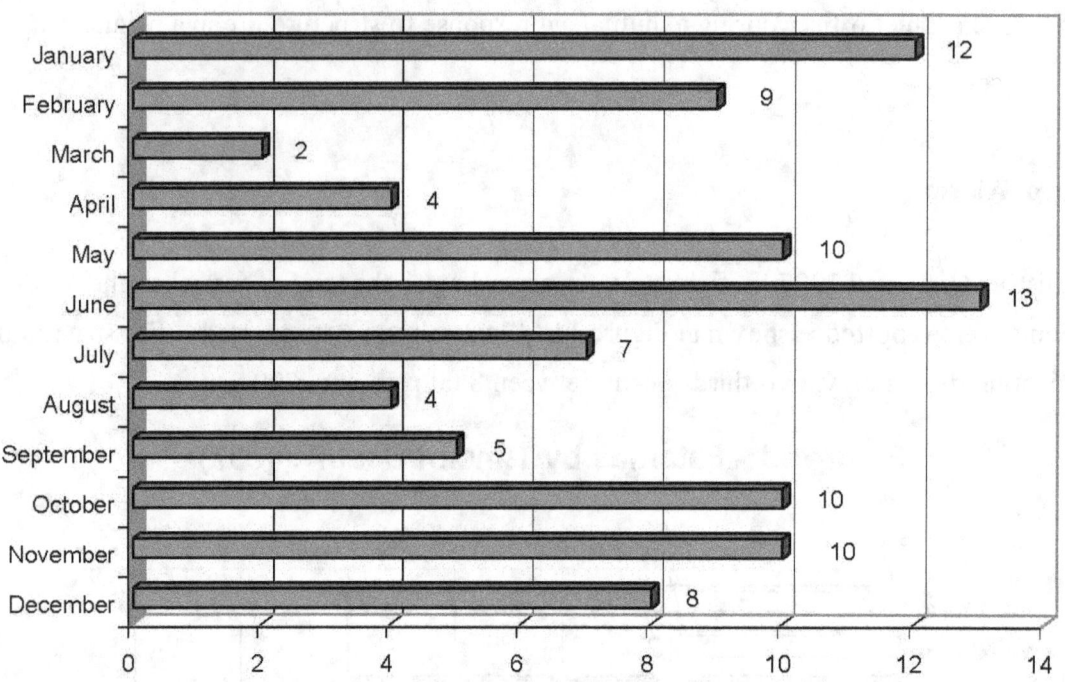

Figure 12 - Deaths by Month of the Year

State and Region

The distribution of firefighter deaths by state is shown in Table 6.[4] Thirty-one states and the District of Columbia had at least one firefighter fatality. New York had the highest number of deaths with 18 followed by Pennsylvania with 12. Figure 13 shows the firefighter fatalities divided by region of the country and their status as career, volunteer, or wildland firefighters.

Figure 13.
Firefighter Deaths By Region 1997

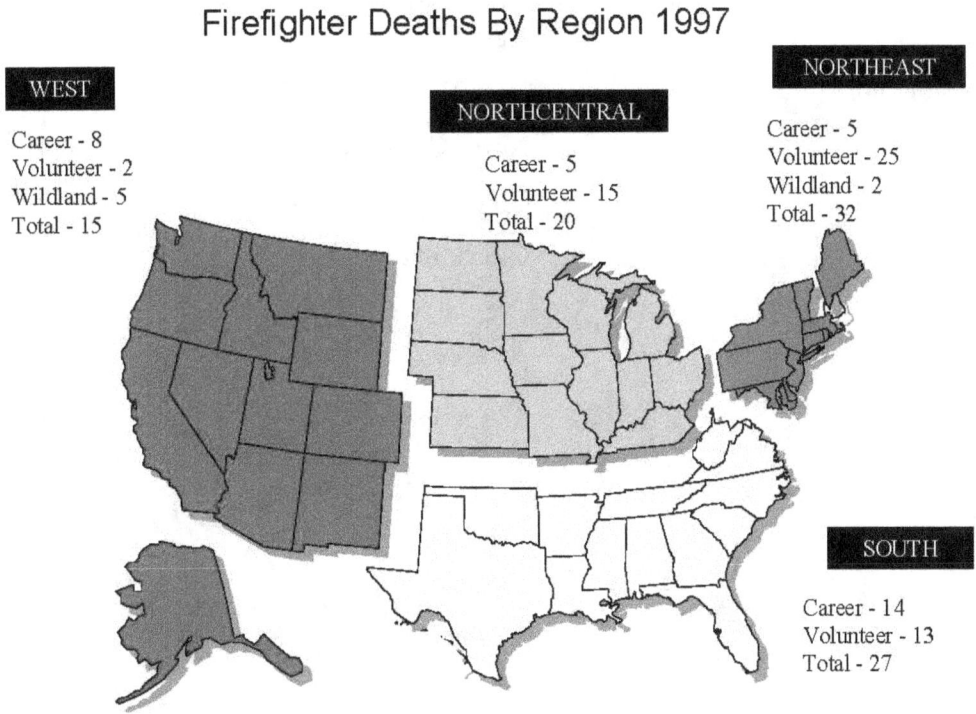

WEST
Career - 8
Volunteer - 2
Wildland - 5
Total - 15

NORTHCENTRAL
Career - 5
Volunteer - 15
Total - 20

NORTHEAST
Career - 5
Volunteer - 25
Wildland - 2
Total - 32

SOUTH
Career - 14
Volunteer - 13
Total - 27

---

[4] This list attributes the deaths according to the state where the fire department or unit is based, as opposed to the state where the death occurred. They are listed by those states for statistical purposes, and for the National Fallen Firefighter Memorial at the National Fire Academy.

Table 6.

1997 State with On-Duty Firefighter Fatalities

State	Number of Deaths	State	Number of Deaths
Alabama	3	Missouri	3
Arizona	2	Nevada	1
Arkansas	3	New Jersey	1
California	9	New Mexico	1
District of Columbia	1	New York	18
Delaware	1	North Carolina	1
Florida	2	Ohio	4
Georgia	2	Oklahoma	1
Illinois	6	Oregon	1
Indiana	1	Pennsylvania	12
Kentucky	2	Rhode Island	1
Louisiana	2	Tennessee	3
Maryland	1	Texas	1
Michigan	2	Virginia	2
Minnesota	2	Washington	1
Mississippi	2	Wisconsin	2

Total - 94

Analysis of Urban/Rural/Suburban Patterns in Firefighter Fatalities

The US Bureau of the Census defines "urban" as a place having a population of at least 2,500 or lying within a designated urban area. Rural is defined as any community that is not urban. Suburban is not a census term but may be taken to refer to any place, urban or rural, that lies within a metropolitan area defined by the Census Bureau, but not within one of the central cities of that metropolitan area.

Fire department areas of responsibility do not always conform to the boundaries used for the census. For example, fire departments organized by counties or special fire protection districts may have both urban and rural coverage areas. In such cases, it may not be possible to characterize the entire coverage area of the fire department as rural or urban, and firefighter deaths were listed as urban or rural based on the particular community or location in which the fatality occurred.

The following patterns were found for 1997 firefighter fatalities. These are estimates based upon population and area served reported by the fire departments.

Table 7.

	Urban/Suburban	Rural	Federal or State Parks/Wildland	Total
Firefighter Deaths	59	27	8	94

SPECIAL TOPICS

FIREFIGHTERS STRUCK BY MOTOR VEHICLES

In 1997, five firefighters were killed as a result of being struck by a motor vehicle while working on the scene of an emergency incident. Four of these incidents occurred as firefighters worked at the scene of motor vehicle collisions and one at the scene of a car fire. Three firefighters died in similar situations in 1996, and data available at the time of this report for 1998 includes at least three such incidents. One incident in 1996, one incident in 1997, and two incidents in 1998, involved the death of a Fire Police member.

Firefighters are often called upon to work in close proximity to moving traffic. Many, if not most, fire departments provide treatment for citizens injured in traffic collisions. At fires, pump operator/engineers are exposed to the risk of being struck due to the fact that the pump panels on most fire apparatus are located on the left side of the apparatus.

These incidents pose a serious threat to firefighters. Some measure of protection can be afforded by the following suggestions:

- **Use apparatus as a shield between the incident scene and the flow of traffic.** The size and sheer bulk of a pumper or ladder truck afford protection from some collisions.

- **Shut down the roadway.** The risks to firefighters posed by passing traffic can be almost completely eliminated by closing the street or road on which the incident has occurred. This alternative is not always popular with law enforcement officials, especially on roads with high traffic volumes and during morning and afternoon commutes.

- **Use Traffic Cones.** Traffic cones are an effective means of channeling traffic away from rescue workers. Flares provide a similar means of traffic channeling but present an ignition hazard. Traffic cones that incorporate retro-reflective bands and florescent colors provide heightened visibility.

- **Illuminate the Scene.** For incidents that occur in twilight, night, near dawn, or in low light situations caused by weather, illuminate the scene. Drivers that can see what hazards are ahead will likely use more caution around incident scenes.

- **Wear clothing that incorporates retro-reflective and florescent fabrics or bands.** Structural protective clothing provides proper protection and visibility for fire incidents but may limit mobility in situations, such as vehicle collisions, when firefighters must work in tight places. Structural protective clothing may also be undesirable for some non-fire situations due to heat stress. Traffic vests with appropriate reflective material, such as those used by highway workers, offer an alternative to structural protective clothing that offers less heat stress potential.

- **Utilize local law enforcement and trained fire police for traffic control.** The positive impact gained with the assistance of trained law enforcement professionals in the control of traffic cannot be overstated. Fire Police, where allowed by law and available, can also provide excellent support and should be trained in proper traffic control procedures and safety.

- **Turn off some emergency lights.** Recent changes to NFPA fire apparatus standards have incorporated a "blocking" mode for emergency lights. This mode provides adequate lighting to identify an emergency vehicle but disallows the use of clear (white) lights which may blind oncoming drivers when the apparatus is parked.

- **Use amber lenses for blocking situations.** The most recent NFPA apparatus standards also recognize the value of amber lights in the blocking mode. Amber lights provide oncoming drivers with sufficient notice of the presence of a hazard while minimizing the "rubbernecking" that comes with the use of red, blue, and clear (white) lights.

- **Manage the incident scene.** Given the hazardous nature of incidents in proximity to traffic, incident commanders may consider limiting on-scene personnel to a minimum while keeping sufficient personnel on scene to perform incident tasks safely. Firefighters and support personnel who are not engaged in necessary activities should be removed from the incident scene. Civilian onlookers should be kept at a safe distance. Fireline tape is an excellent means to corral and control onlookers, as is a law enforcement presence.

VEHICLE ACCIDENTS

In 1997, fourteen firefighters were killed in accidents (vehicle collisions) while responding to or returning from an incident. This number includes one firefighter who was killed when the ambulance in which he was riding lost control on wet pavement and overturned. This number does not include the deaths of firefighters who were killed in crashes of wildland fire fighting aircraft.

For every year since 1993, vehicle accidents have been the second leading cause of death for firefighters. In 1995, thirty firefighters lost their lives in vehicle accidents, the toll dropped to twenty-two in 1996, and finally to fourteen in 1997. Although the number of deaths has dropped over the last few years, there is still reason for concern and a need to continue efforts to lower these incidents even further.

A disproportionate number of these deaths involve volunteer firefighters, many occur in their personal vehicle. In 1997, twelve of the fourteen firefighters killed in vehicle collisions were volunteers. Five of these volunteer firefighter deaths involved personal vehicle accidents including one incident which might be considered an act of nature where a tree fell on a firefighter's car while he was enroute to an emergency call. In 1996, eight volunteer firefighters died as the result of vehicle accidents while operating their personal vehicles

One of most significant risks faced by firefighters is the risk of injury and death while responding to emergencies. Every fire department should have a plan in place to manage these risks. The plan should be specific to the needs and operations of the department and should include information on driver qualifications, driver training, requirements for the use of lights and sirens for response, policies on safe speed for apparatus operation, policies on requirements for negative right of way situations (such as stop signs, red lights, and response in opposing lanes of traffic), and requirements for the use of a backup spotter. Fire departments that allow members to respond in their personal vehicles may need to pay special attention to training their members in safe response modes.

Recently, fire departments have begun to classify some responses, or some parts of the response complement to an incident, as non-emergency or "on the quiet". Examples of non-emergency response include sending firefighters to less severe incidents without the use of lights and siren; sending the closest unit to the emergency with lights and siren while the rest of the assignment proceeds without their use; or having the first arriving unit shut the rest of the response down to a non-emergency mode after performing an initial on-scene assessment of the problem. A response risk reduction option employed by some volunteer fire departments involves a crew of firefighters sleeping at the fire station and handling minor emergencies without alerting the entire fire department. Any effort to minimize emergency mode response and the risks associated with it can have a positive impact on firefighter injuries and deaths.

A publication available from the USFA entitled *"Alive on Arrival"* provides excellent information on safety during response. The publication (item number 5-0195, publication ID number 186) is free and can be obtained through the internet at **http://www.usfa.fema.gov/usfapubs** or by writing:

> United States Fire Administration
> Publications Center
> 16825 South Seton Avenue
> Emmitsburg, MD 21727

CONCLUSIONS

The analysis of firefighter deaths in 1997 indicates that the overall long-term trend toward fewer firefighter fatalities is continuing. The 94 fatalities in 1997 are the third lowest recorded since the inception of this study, and only the fifth time the total number of fatalities has dropped below 100. Five of the last six years have resulted in less than 100 firefighter fatalities. Firefighters die while engaged in a wide range of services to the community. The addition of emergency medical services provision, hazardous materials response, and technical rescue service has brought with them risks and accompanying firefighter deaths.

Although the long term reduction in the number of deaths is reason for hope, firefighters are still dying at an unacceptable rate.

Stress-induced heart attacks remained the top cause of firefighter deaths. Continued focus on firefighter health and wellness may reduce the impact of this killer in the future. There have been a number of studies of this issue including the recent Wellness Initiative between the International Association of Fire Chiefs (IAFC) and the International Association of Fire Fighters (IAFF). This document provides a framework for long term attention to issues of firefighter health and wellness. Copies of this report are available for sale from the IAFC. Copies are also available to IAFF members from the IAFF. A cost may be involved.

International Association of Fire Chiefs
4025 Fair Ridge Drive
Fairfax, VA 22033-2868
http://www.iafc.org

International Association of Fire Fighters
1750 New York Avenue, NW
Washington, DC 20006
http://www.iaff.org

The second leading activity at the time of death for 1997 was vehicle accidents. Many of these deaths were preventable. The development and management of procedures for emergency responses could contribute greatly to a reduction in deaths. Seatbelts were not worn in at least four incidents and may have changed the outcome of some of these incidents. Five deaths involving flight crews of wildland fire fighting aircraft remind us of the hazards associated with their deployment.

Two firefighter deaths occurred in 1997 as the result of injuries and exposures that occurred in previous years. While their deaths are not included in this analysis, these firefighters will be included in the 1997 annual Fallen Firefighter Memorial Service at the National Fire Academy, and their names will be included on the list of firefighters who died in 1997.

Ten firefighters died of asphyxiation in structure fires. An operating PASS device may not have saved all of them but would have likely changed the outcome of some of these incidents. In a several cases, PASS devices were worn but not activated. Most SCBA manufacturers offer integrated PASS devices that activate when the SCBA is utilized. Kits are available from some manufacturers to retrofit existing SCBA's with this life saving device.

APPENDIX A

SUMMARY OF 1997 INCIDENTS

1/1/97
Brian D. Myers, Sr., Engineer
Age 47, Volunteer
Schuylerville Hose Company, NY
Firefighter Myers and three other firefighters were operating the nozzle at the scene of a restaurant fire early on New Year's Day when the ceiling collapsed and a flashover occurred. Two of the firefighters were able to escape. A rescue team placed a nozzle through the front window to cool down the area where firefighters were trapped. They located and removed one of the trapped firefighters. They then reentered the building and found Firefighter Myers. He too was removed and emergency medical care was provided. Firefighter Myers went into cardiac arrest while enroute to the hospital. He had suffered burns to 70 percent of his body. His son and one other firefighter were also injured.

1/2/97
David P. Janora, Assistant Chief
Clarence Center Fire Department, NY
Age 49, Volunteer
Chief Janora went into cardiac arrest during a meeting at the fire station. Chief Janora had attended a vehicle fire earlier during the day. Chief Janora was afflicted at 9:30 p.m.

1/2/97
Harold "Mac" E. McGowan, Firefighter/Safety Officer
Age 70, Volunteer
Union Fire Company #1, Lebanon City Fire Department, Lebanon, PA
Firefighter McGowan went into cardiac arrest immediately following a structure fire. At the end of the fire emergency, Firefighter McGowan removed his gear and collapsed.

1/3/97
Arthur R. Ebert, Firefighter
Fort Morrow Fire Department, Waldo, OH
Age 63, Volunteer
Firefighter Ebert had responded to one structure fire and returned to the station to clean up, when they were dispatched to a report of a house fire. Firefighter Ebert and another firefighter took an engine to the scene. They were the first to arrive. The other firefighter pulled a handline, but was not getting any water. He went to see why and found Firefighter Ebert had collapsed due to a heart attack.

1/7/97
Carl L. Ayers, Fire Police
Newton-Ransom Fire Company, Clarks Summit, PA
Age 67, Volunteer
Firefighter Ayers was struck by a car and killed while directing traffic at a motor vehicle collision.

1/8/97
H. Robert Hathaway, Chief
Branchport Fire Department, NY
Age 58, Volunteer
Chief Hathaway collapsed due to a heart attack immediately following a meeting at the fire house.

1/10/97
Harold Hester, Firefighter
Malden Fire Department, MO
Age 52, Volunteer
Firefighter Hester was involved in a vehicle accident while responding to a call. He was in his personal vehicle when the car hit a patch of ice on the highway and lost control.

1/10/97
Allen H. Martin, Jr., Firefighter
New Orleans Fire Department, LA
Age 33, Career
Firefighter Martin became trapped by debris in a two-story residential structure fire while searching the structure for victims and conducting an interior attack. The roof collapsed and it was several minutes before firefighters were able to make a rescue attempt. Firefighter Martin died shortly after arriving at the hospital.

1/14/97
Stoy Geary, Chief
Rosine Volunteer Fire Department, KY
Age 62, Volunteer
Chief Geary went into cardiac arrest at the scene of a residential structure fire.

1/15/97
Richard Sanders, Lieutenant
Oakland Fire Department, CA
Age 47, Career
Lieutenant Sanders died as a result of a heart attack that occurred early in the morning in the station towards the end of his shift.

1/26/97
Robert William Martinson, Assistant Chief
Conover Fire Department, WI
Age 43, Volunteer
Chief Martinson was on the roof of a house checking the progress of the fire when the roof collapsed. Chief Martinson succumbed to smoke inhalation.

2/4/97
Wayne M. Fogel, Firefighter
Detroit Fire Department, MI
Age Unknown, Career
Firefighter Fogel suffered from a heart attack while on-duty at the fire house.

2/5/97
Kevin C. Seaburg, Assistant Chief
Selkirk Fire District, NY
Age 38, Volunteer
Chief Seaburg collapsed due to a heart attack at the scene of a structure fire while carrying and setting up equipment.

2/6/97
Bryan J. Golden, Firefighter, Age 21 - Career
Brett A. Laws, Firefighter, Age 29, Career
Stockton Fire Department, CA
Units were dispatched to a report of a house fire. The first arriving officer found a working fire and immediately requested a second alarm assignment. Two houses were on fire and there was a possibility of a person trapped. Unbeknownst to the initial crews, the house was much bigger than it appeared from the street and there was a large two story addition heavily involved in fire. An interior attack was initiated with a 1-3/4 inch handline through the front door.

Approximately 21 minutes later, with no warning, there was a catastrophic collapse of the entire second floor and roof of the addition. The collapse trapped firefighters working on the first floor. Fire Captain Oscar Barrera was trapped in the burning debris, but was rescued through the heroic efforts of other firefighters. Captain Barrera was seriously burned. Firefighter Laws and Firefighter Golden were killed and could not be rescued.

The owner of the house was also killed in the fire and was found later after the fire was extinguished. The second story had been added on by the owner and was made of heavy timber. It had been used as a dance studio. This was Firefighter Golden's first fire.

2/15/97
Timothy J. Warren, III, Firefighter
Geneva Fire Department, NY
Age 36, Volunteer
Firefighter Warren collapsed due to a heart attack while fighting a fire that had broken out in a three-story dormitory at Hobart College. The fire started on the first floor. No students were injured.

2/16/97
Peter Kahn, Firefighter
Trumansburg Fire Department, NY
Age 75, Volunteer
Firefighter Kahn collapsed due to a heart attack while directing traffic at the scene of a two-car vehicle collision.

2/17/97
Charles "Chuck" H. Williams, II, Firefighter
Lexington Fire Department, KY
Age 29, Career
Firefighter Williams and a second firefighter became trapped after entering a residential fire and falling through a hole into the basement. Both received second and third degree burns. Efforts were made to revive Williams on the scene. The other firefighter was admitted to the hospital with serious burns.

2/22/97
Robert E. Fowler, Firefighter
Spencerport Fire Department, NY
Age 54, Volunteer
Firefighter Fowler was crushed and killed when a tree fell on his personal vehicle during a response to an emergency call. Due to extreme weather, the fire department had been called out on several downed power lines and windows that had been blown out. Firefighter Fowler and his son were driving the half mile from their house to the station. His son, who was a junior firefighter, was also taken to the hospital with back injuries.

2/28/97
Charles Allen Weber, Sr., President
Violetville Volunteer Fire Company, Baltimore County, MD
Age 48, Volunteer
Firefighter Weber collapsed due to a heart attack in quarters immediately after having returned from a call. After returning to the station, he began storing his gear and collapsed.

3/15/97
Russett "Rusty" S. Hauber, Firefighter/Dive Team
Yakima Fire Department, WA
Age 32, Career
Firefighter Hauber died from asphyxiation while attempting a rescue of two civilian divers from a 1,000 foot long, 100 foot deep irrigation siphon. A police officer was also killed in association with this incident.

3/21/97
Tommy T. Gross, III, Firefighter
Tuscaloosa Fire Department, AL
Age 24, Career
Firefighter Gross suffered from a heart attack while going through a burn building while in rookie school.

4/17/97
Larry L. Mercer, Captain
Duncan Fire Department, OK
Age 48, Career
Captain Mercer died as a result of a stroke that occurred while on-duty. He was on a 24 hour shift and was feeling ill. This prompted him to make a doctor's appointment for the next working day. He died as a result of the stroke 48 hours after experiencing pain on his way to work.

4/20/97
William W. Babka, Firefighter/Co-Pilot, Age 34 - Career
Walter John Hirth, Jr., Captain/Pilot, Age 45 - Career
Pennsylvania Bureau of Forestry, Harrisburg, PA
Pilot John Hirth and Co-Pilot William W. Babka were killed after approaching a forest fire, started by an unattended campfire, to make a water drop. After leveling for the drop, the aircraft was affected by a downdraft from wind gusts. The smoke from the fire also had an impact on their visibility. The aircraft stripped off the tops of trees for approximately 100 feet before coming to rest.

4/26/97
Earl Holsapple, Captain
California Division of Forestry, La Cima Fire Center, Julian, CA
Age 45, Volunteer
Captain Holsapple died while teaching an equipment operating class at the California Division of Forestry's Fire Academy. Captain Holsapple suffered from a severe asthma attack that led to cardiac arrest.

5/3/97
Jessie F. Bricker, Jr., Fire Apparatus Operator
San Antonio Fire Department, TX
Age 47, Career
Fire Apparatus Operator Bricker died as a result of cardiac arrest that was exacerbated by stress and smoke inhalation that occurred while fighting a fourth alarm motel fire. After returning from the fire, FAO Bricker complained of not feeling well. He was transported to the hospital that morning at 5:56 a.m. in severe respiratory distress. He died at approximately 4:38 p.m. Other personnel also suffered from various symptoms as a result of exposure of the fire and smoke. The following day, an environmental consultant was directed to sample and analyze the scene and clothing worn by firefighters. Analysis revealed that an "unusual chemical event" occurred at the scene of the fire. Testing determined that chemicals that firefighters may have been exposed to included chlorine or chlorine compounds, hydrochloric acid, pesticides, amines, illicit drugs (such as methamphetamines) and other chemicals not determined.

5/3/97
Tracy D. Floyd, Firefighter
Winchester Fire Department, TN
Age 29, Career
Firefighter Floyd was killed while responding to the scene of a structure fire when another vehicle pulled out in front of him, causing a collision.

5/8/97
M. Edward Hudson, Lieutenant, Age 53 - Career
Reginald G. Robinson, Sr., Firefighter, Age 33 - Volunteer
Stewart Warren, Captain, Age 47 - Career
West Helena Fire Department, AR
The West Helena Fire Department was dispatched to the BPS Bartlo chemical plant at 1:02 p.m. on May 8, 1997. They were informed that there was smoke coming from the building and that it contained Azinphos Methyl. The building exploded at approximately 1:22 p.m., killing three firefighters and severely injuring one. Eleven other firefighters were involved in the rescue of the injured firefighter and the rescue attempt for the three firefighters who died.

5/9/97
Willie E. Rowe, Jr., Captain
Macon-Bibb County Fire Department, Macon, GA
Age 49, Career
During a severe thunderstorm, Captain Rowe and his crew responded to a report of a trash fire and found a downed tree on the roof of a house. Captain Rowe and another firefighter went behind the house to assess the situation when the tree began to slide off the roof. The firefighters began to move out of the way when Captain Rowe slipped in the mud and was crushed by the falling tree. He was killed instantly.

5/12/97
Lawrence Hobson, Lieutenant
Robbins Fire Department, IL
Age 49, Career
Lieutenant Hobson collapsed and died due to a heart attack while pulling 2-1/2 inch hoseline at an abandoned house fire.

5/24/97
Timothy M. Goff, Firefighter
Kenmore Volunteer Fire Department, NY
Age 27, Volunteer
Firefighter Goff died on May 24, 1997, as a result of injuries sustained from a wall collapse during a paint store fire on May 5, 1997. Five other firefighters were injured when the wall collapsed.

5/25/97
William "Junior" T. Wilson, Firefighter
Pinecrest Volunteer Fire Department, Jacksboro, TN
Age 41, Volunteer
Firefighter Wilson was killed during a rollover enroute to a vehicle accident on Interstate 75.

5/26/97
Stanley F. Kaminski, Firefighter
Langford-New Oregon Fire Department, North Collins, NY
Age 67, Volunteer
Firefighter Kaminski went into cardiac arrest and died while attending a memorial service with the fire department.

5/29/97
David M. Ray, Firefighter
California Department of Forestry - Conservation Corps - Julian, CA
Age 21, Contract
Firefighter Ray died as a result of a heat stroke after fighting a brush fire. The fire started when a tractor mower hit a rock, producing a spark. Over 140 firefighters fought the fire. Two firefighters went to the hospital after collapsing from heat stroke. According to news articles, this was Firefighter Ray's first fire. Firefighter Ray's crew was assigned to cut line on a hillside. Winds at the time were light and Westerly. The temperature was close to 100 degrees Fahrenheit by early afternoon, and the humidity level was under 15%. After a long period on line cutting (which included one break), Firefighter Ray manifested symptoms of heat stress. Despite the immediate on-scene attention of his crew and quick evacuation to a hospital, his condition deteriorated rapidly to heat stroke. He did not regain consciousness and died early the next morning.

6/3/97
Jesse Gates, Pilot, Age Unknown - Career
Leo A. Stevens, Firefighter, Age 55 - Career
Fort Apache Indian Reservation and BIA Facility Management, Whiteriver, AZ
Pilot Jesse Gates and Firefighter Leo Stevens were killed in the crash of a fire reconnaissance airplane on the Fort Apache Indian Reservation in Eastern Arizona. A lookout tower reported black smoke near the Black River. Dispatch lost contact with the patrol plane at about the same time the smoke was reported. The plane had been on a routine fire reconnaissance flight.

6/5/97
James H. Johnson, Firefighter , Age 63 - Volunteer
Charles A. Rudd, Lieutenant, Age 21 - Volunteer
New Bloomington Fire Department, OH
Firefighter Johnson and Lieutenant Rudd were electrocuted at the scene of a motor vehicle collision when a rescuer came into contact with downed power lines, creating a chain reaction. Five rescuers were injured and three rescuers were electrocuted. The initial victim, who was on a backboard being carried, was also electrocuted. Rural/Metro paramedic Robert Good was also killed at this incident.

6/7/97
Gerald T. Ertle, Captain
Benton Volunteer Fire Department, MI
Age 53, Volunteer
Captain Ertle went into cardiac arrest and died during a training class at the state academy. After completing certification classes during April, May, and June, he attended the volunteer certification field day (practical testing day) at the fire academy. Shortly after completing the testing, Captain Ertle suffered a fatal heart attack.

6/9/97
Timothy Wayne Martin, Firefighter/EMT
Clovis Fire Department, NM
Age 38, Career
Firefighter Martin was providing patient care in the back of an ambulance when the ambulance lost control on wet pavement and overturned. The patient was also killed.

6/16/97
Edwin J. Haungs, Sr., Fire Police
South Lockport Fire Company, Inc., Lockport, NY
Age 51, Volunteer
Fire Police Officer Haungs' fire company was called to assist with traffic control at a mutual aid motor vehicle accident. He collapsed and died due to a heart attack immediately after returning from this call.

6/16/97
William Jack Northam, Firefighter
Laurel Fire Department, Inc., DE
Age 55, Volunteer
The Laurel Fire Department was dispatched to a motor vehicle accident. Firefighter Northam was putting away tools before boarding the apparatus when he collapsed due to a heart attack. He was taken to the hospital and died ten hours later.

6/16/97
John McClay Watson, Firefighter
Moscow Volunteer Fire Company, PA
Firefighter Watson died as a result of injuries sustained from an accident involving his personal vehicle while enroute to an EMS call. He was thrown from his vehicle, which rolled over three times in the course of the accident. He died the next day as a result of severe head trauma.

6/19/97
William C. Mellon, President
Bay Ridge Volunteer Fire Department, Lake George, NY
Age 58, Volunteer
Firefighter Mellon suffered from a heart attack while preparing to respond in his POV to a fire alarm.

6/20/97
Michael F. Drobitsch, Firefighter
Chicago Fire Department, IL
Age 46, Career
Firefighter Drobitsch's death occurred during a diving training session.

6/22/97
Michael E. Neuner, Sr., Lieutenant
Brewster Fire Department, NY
Age 35, Volunteer
Lieutenant Neuner died due to injuries sustained at a residential structure fire after becoming trapped in the basement.

6/28/97
Ricky G. Moore, Firefighter
Oak Grove/Thach Fire Department, Toney, AL
Age 29, Volunteer
Firefighter Moore was killed when he was thrown from a fire truck after it overturned enroute to a fire call.

7/3/97
Joseph M. Vagnier, Firefighter/EMT
Monroeville Volunteer Fire Company #4, PA
Age 21, Volunteer
Firefighter Vagnier and two other firefighters were attempting to rescue a flood victim when Vagnier was swept under the wheel of a truck by raging waters. Firefighters found Vagnier, treated him at the scene, and immediately transported him to a local hospital where he subsequently died. Reports indicate that either Vagnier was pulled under when another firefighter holding his security rope went under or he was pulled under after attempting to help rescue another firefighter that went down. The incident occurred on 7/1/97, Firefighter/EMT Vagnier died on 7/3/97.

7/4/97
Michael L. Seguin, Firefighter
Buffalo Fire Department, NY
Age 31, Career
Firefighter Sequin was killed when he became trapped by a roof collapse while fighting a residential structure fire. One other firefighter was injured and suffered second degree burns. The second firefighter was dragged to safety after becoming unconscious. Rescuers did not see Firefighter Sequin due to heavy smoke and he was not located until later that afternoon. Fire officials stated that there was a possibility that the fire was started by fireworks. The owner of the house believes that a "rocket" landed on her roof.

7/6/97
Dean Sierra Hiser, Sr., Pilot
Sierra National Forest, Clovis, CA
Age Unknown, Wildland Contract
Pilot Hiser was killed as a result of a helicopter crash that occurred during a water drop at a wildland fire.

7/13/97
James H. Tebo, Captain
Ranger Community Fire Department, Bangor, MI
Age 61, Volunteer
Captain Tebo collapsed at the scene of a structure fire and later died as a result of a heart attack.

7/15/97
Malcolm A. Rovero, Firefighter
Estero Fire Protection and Rescue District, FL
Age 34, Career
Firefighter Rovero died from apparent cardiac and respiratory arrest at the scene of a ten acre brush fire.

7/16/97
Albert Sipple, Lieutenant
Bellmore Fire Department, NY
Age 50, Volunteer
Lieutenant Sipple collapsed of a seizure while assisting a victim at a motor vehicle accident.

7/25/97
Jerome H. Chlian, Jr., Firefighter
Starbuck Fire Department, MN
Age 46, Volunteer
Firefighter Chlian died as a result of a heart attack while on-duty.

8/3/97
Joseph J. Estavilla, Fire Engineer
San Diego Fire & Life Safety Services, CA
Age 44, Career
Firefighter Estavilla was called out on a strike team to fight a brush fire in the Northern part of San Diego County. While fighting the fire, Estavilla sustained cuts on his hands (through his gloves). After returning to his crew that night (2:00 a.m.), Estavilla complained to his Captain about not feeling well (this was not like him). They immediately went to the hospital. By morning they had discovered that Firefighter Estavilla was infected with Strep A. This disease spreads in 24-48 hours with symptoms similar to the flu. Firefighter Estavilla died on August 3rd.

8/19/97
Jeffrey E Sammons, Firefighter
South Whitley Fire Department, IN
Age 21, Volunteer
Firefighter Sammons was killed and two others were injured in a restaurant fire caused by cooking equipment that had been left on. Sammons and others were making an internal fire attack when the heat buildup became extreme. They started to exit the structure when a flashover occurred causing some of the ceiling tile to fall.

8/22/97
Richard B. Jenkins, Sr., Firefighter
Tennville Fire Department, GA
Age 30, Volunteer
Firefighter Jenkins was killed enroute a to house fire from his home in a private vehicle.

8/31/97
Robert D. Chisholm, Assistant Chief
Gearhart Fire Department, OR
Age 50, Volunteer
Chief Chisholm had a heart attack while trying to rescue a missing swimmer at Gearhart beach in Oregon. After searching for the victim, Chisholm became tired and passed out. His fellow firefighters dragged him to the shore. Crews were unable to revive Chisholm.

9/7/97
Howard E. Strube, Firefighter
Canton Fire Department, IL
Age 34, Career
Firefighter Strube was killed during training on the fire department's new aerial apparatus. Firefighter Strube was operating as the safety observer (platform operation) when his head became caught between the ladder rungs while the ladder was retracted.

9/8/97
David E. Carpenter, Firefighter, Age 38 - Volunteer
Donald J. Payton, Sr., Captain, Age 57 - Volunteer
Thayer Rural Fire Department, MO
Captain Payton and Firefighter Carpenter were both killed while responding to a motor vehicle accident. The fire truck they occupied collided head on with a dump truck. The police reported that the dump truck crossed the center line and struck the fire truck.

9/9/97
Kenneth E. Bayer, Captain
Los Angeles County Fire Department, CA
Age 52, Career
Captain Bayer died on September 9, 1997, of cardiac arrest. On September 5, 1997, he was exposed to high concentrations of carbon monoxide (CO). The exposure to smoke and CO occurred over an approximate 45-60 minute period as he directed the interior extinguishment, salvage, and overhaul operations of a chimney and attic fire in a two story condominium.

9/14/97
Henry E. Perry, Firefighter
Pumpkin Center Fire Department, Inc., Jacksonville, NC
Age 59, Volunteer
Firefighter Perry went into cardiac arrest at the scene of a structure fire. He was climbing on top of a fire truck when he fell. The pump operator noticed him falling, but had no time to react. Firefighter Perry died as a result of injuries received by the fall.

10/2/97
W. Douglas Buckert, Firefighter, Age 23 - Volunteer
Michael D. Mapes, Firefighter, Age 35 - Volunteer
Carthage Fire Department, IL
These fatalities occurred at a fire located at a grain dryer fire North of Burnside, IL. Enroute to the fire, firefighters were advised that LPG tanks were involved. Upon arrival, firefighters noticed one of the 1000 gallon tanks venting and shooting flames approximately 45 to 50 feet in the air. After surveying the scene and talking to the owner, it was decided to move the truck to a safer location for a better point of attack. While doing so, one of the tanks exploded, causing the deaths of Firefighter Mapes and Firefighter Buckert. Two other firefighters were injured.

10/10/97
William "Pops" H. Winters, Deputy Chief
Atglen Fire Company, PA
Age 76, Volunteer
Chief Winters collapsed and went into cardiac arrest at the scene of a structure fire.

10/15/97
Harold "Ray" Elliott, Battalion Chief
Kern County Fire Department, CA
Age 54, Career
Chief Elliott was doing mandatory PT outside at Virginia Colony Station (41) when he collapsed due to a heart attack. He was on an overtime shift and had already worked approximately 24 hours. He collapsed at 9:44 p.m. Chief Elliott had 32 years on the job when he collapsed. Chief Elliott died on April 28, 1997.

10/22/97
David S. Williams, Firefighter
Perry Fire Department, FL
Age 26, Career
Firefighter Williams was killed when an overheated bullet (.22 caliber) from a greenhouse fire discharged hitting Williams in the chest.

10/24/97
John M. Carter, Sergeant
District of Columbia Fire Department, Washington, DC
Age 38, Career
Sergeant Carter died when the floor beneath him collapsed during a three alarm grocery store fire in our nation's capitol. The building was being evacuated by the incident commander at the time of the collapse. The investigation into the fire's cause indicated that faulty electrical wiring in the basement started the fire.

10/24/97
David Womer, Firefighter/EMT
Mount Carmel Volunteer Fire Department Station 5, PA
Age 24, Volunteer
Firefighter Womer was killed when the rescue squad building where he was on-duty experienced an explosion. A fellow rescue squad member rolled an open thirty pound propane tank from a gas grill into the squad house as a practical joke. All but Womer evacuated the building. Eventually the gas reached the pilot light in the furnace room and triggered the explosion.

10/26/97
Kathryn A. Mayfield, Firefighter/EMT
Crooksville Volunteer Fire Department, OH
Age 47, Volunteer
Firefighter Mayfield collapsed in the station after returning from a tire fire and was taken to the hospital. She died the next morning due to a heart attack. Firefighter Mayfield was stricken on 10/25/97 and died on 10/26/97.

10/27/97
James E. Hynes, Firefighter, Age 27 - Career
Terry McElveen, Lieutenant, Age 43 - Career
Philadelphia Fire Department, PA
Lieutenant McElveen and Firefighter Hynes died as a result of smoke inhalation at the scene of a residential structure fire. The fire was a result of wires that had come down on the roof during a heavy rain. The firefighters were operating in the interior of a two-story occupied dwelling with a fire in the basement. They both ran out of air, removed their SCBA masks, and remained inside the dwelling. The two firefighters were found near the back door with their SCBA's on, but their masks off.

11/2/97
Leroy Swenson, Captain
Minneapolis Fire Department, MN
Age 56, Career
Captain Swenson was killed at the scene of a four-vehicle accident when a large commercial truck lost control on the icy road and rolled over on top of Swenson killing him instantly.

11/5/97
William S. Bradner, III, Firefighter
Tunstall Fire and Rescue Company, Danville, VA
Age 30, Volunteer
Firefighter Bradner was killed when he was thrown from a tanker truck as it overturned returning to the scene of a structure fire. Another volunteer was also injured in the accident and suffered a broken pelvis. The two firefighters had already delivered one load of water and were returning with the second load when the accident occurred.

11/6/97

Johnson "Jack" Oatman, Firefighter

Ewansville Volunteer Fire Department, Mount Holly, NJ

Age 55, Volunteer

Firefighter Oatman died from a heart attack that occurred while preparing to respond to a motor vehicle accident with entrapment.

11/8/97

John F. Kroening, Firefighter

Cambria Fire Company, Lockport, NY

Age 75, Volunteer

Firefighter Kroening was on the scene of a motor vehicle accident setting up equipment when he was struck by a passing vehicle. He was taken to the hospital and died from the injuries the next day. The incident occurred on 11/7/97 and Firefighter Kroening died on 11/8/97.

11/13/97

Eugene Ottonello, Assistant Fire Manager Officer

Bureau of Land Management, Battle Mountain, NV

Age 47, Wildland Career

Officer Ottonello died due to an asthma attack at a prescribed burn. Respiratory arrest led to heart failure. Crews initiated CPR, but to no avail.

11/14/97

Scott Alan Vrabel, Firefighter

New Salem Volunteer Fire Department, PA

Age 26, Volunteer

Firefighter Vrabel was killed when he lost control of a brush truck enroute to a two vehicle accident. The truck ran off the road into a telephone pole and flipped - pinning Vrabel underneath.

11/15/97

William H. Fairweather, Fire Police

Middletown Fire Department, NY

Age 78, Volunteer

Fire Police Officer Fairweather was directing traffic (fire police) at the scene of a motor vehicle accident, when he had a heart attack and died.

11/22/97

Gregory I. Quinn, Assistant Chief

Village of Westfield Fire Department, WI

Age 46, Volunteer

Chief Quinn died after being hit by a sport utility vehicle that lost control on an icy bridge at the scene of a motor vehicle accident. The car struck two firefighters and hit one of the cars involved in the first wreck. The other firefighter was not seriously injured. The driver of the SUV was not wearing her seat belt and was killed.

11/23/97
George A. Davis, Firefighter 1st Class
Houma Fire Department, LA
Age 27, Career
Firefighter Davis entered a structure fire wearing breathing apparatus. He was seen later exiting the structure, still with an SCBA, when he collapsed and fell into the arms of another firefighter. Firefighters performed CPR on the scene. Davis was then taken to the hospital where he died several days later. Firefighter 1st Class Davis was the victim of a heart attack. This incident occurred on 11/18/97 and Firefighter 1st Class Davis died on 11/23/97.

11/28/97
George Hopey, Jr., Fire Police
Dravosburg Volunteer Fire Company #1, PA
Age 69, Volunteer
Fire Police Officer Hopey had a heart attack and died while directing traffic at the scene of a structure fire. He was stricken on 11/26/97 and died on 11/28/97.

12/1/97
Thomas M. McCormack, Chief
Watervliet Fire Department, NY
Age 44, Career
Chief McCormack died as a result of a heart attack while directing fire operations at a residential structure fire mutual aid call in the City of Troy, NY. The City of Watervliet was first due on the call while Troy units were deployed as a "In progress, second alarm".

12/6/97
John F. Lincoln, Jr., Firefighter
Purcellville Volunteer Fire Department, VA
Age 52, Volunteer
Firefighter Lincoln died due to a heart attack after returning from an all night working residential fire that included a civilian casualty. Firefighter Lincoln was stricken on 12/5/97 and died on 12/6/97.

12/9/97
"Randy" Smartt, Firefighter
Huntsville Fire Department, AL
Age 49, Career
Firefighter Smartt died as a result of a heart attack that occurred while on-duty at the fire house.

12/11/97
Ronald A. Guilmette, Private
Woonsocket Fire Department, RI
Age 38, Career
Private Guilmette responded on Engine 4 to an alarm of fire at 1:30 a.m. Upon returning to the station at 1:40 a.m., Private Guilmette complained of back pain to his crew and told them he thought he would be more comfortable sitting up on the couch in the television room. At 5:30 a.m., Engine 4 was dispatched to another alarm. The fire crew found Private Guilmette collapsed on the floor near the couch.

12/15/97
Leonard N. Zeller, Firefighter
Edwards Fire Department, NY
Age 53, Volunteer
Firefighter Zeller died as a result of a heart attack that occurred while responding to an EMS call. He was on foot on the way to the fire station when he collapsed. He was revived on the way to the hospital, but was pronounced DOA at the hospital.

12/17/97
Scott M. Berry, Firefighter/Driver
Bradley County Volunteer Fire Department, Cleveland, TN
Age 33, Volunteer
Firefighter/Driver Berry was killed when the 12,050 gallon tanker overturned while responding to the scene of a brush fire. His brother was also in the tanker and was transported by helicopter to the hospital. Firefighter Berry's father was killed in the line of duty 30 years ago. Firefighter Berry was wearing his seat belt at the time of the accident.

12/20/97
William "Sam" Smitherman, Sr., Firefighter
East Oktibbeha Fire Department, Starkville, MS
Age 59, Volunteer
Firefighter Smitherman was struck by a car at a motor vehicle fire. He was getting ready to retrieve a crow bar to pop the trunk and was hit.

12/23/97
Brian T. Hauk, Assistant Chief
Logan-Trivoli Fire Department, Hanna City, IL
Age 32, Volunteer
Chief Hauk died in a vehicle accident while responding to the fire house for a reported oven fire in an apartment complex. Chief Hauk was taking evasive action to avoid another vehicle which failed to yield to his vehicle when his vehicle flipped. Chief Hauk was displaying an activated blue light. Chief Hauk had the right of way and the other car had a stop sign.

The following firefighters died in 1997, from injuries or exposures that they received prior to 1997:

Mike Gilberg, Captain
Los Angeles County Fire Department, CA
Age 53, Career
Captain Mike Gilberg died on May 1, 1997 as a result of AIDS contracted through documented needle sticks received when he was on the job as a paramedic. He retired two years prior, when he became too weak to work

Frank Delano Gilbert, Jr., Firefighter-Paramedic
Portage Fire Department, IN
Age 39, Career

Firefighter-Paramedic Gilbert died on May 29, 1997, as a result of injuries he received in a collision between an ambulance and a truck. Firefighter Gilbert was part of a crew transporting a drug overdose patient when a truck that had appeared to be yielding the right of way to the ambulance pulled out, hitting the ambulance and causing it to strike a tree.